The war on
slugs and snails
starts here!

Gardening **organically**

One of the great joys of gardening is to experience the variety of life that a healthy garden contains. A garden managed using organic methods will have far more interest in it than a garden where insecticides and chemicals are used. An organic garden is a more balanced environment, where 'good' creatures such as ladybirds and beetles keep the 'bad' pests and diseases under control.

Organically grown plants also tend to be healthier and stronger than plants that rely on large doses of artificial fertiliser. In healthy soil they grow strong roots and can better withstand attack by pests and diseases. Soil can be kept in top condition by recycling garden waste to make nutritious compost. Growing the right combination of plants in the right place at the right time – by rotating where you plant your veg for example, or choosing shrubs to suit the growing conditions that your garden can offer – can deliver impressive disease-free results.

These are the basic principles of organic growing – use the natural resources you already have to create a balanced and vibrant garden. It's sustainable, cheaper than buying chemicals, easier than you think and great fun. Enjoy your organic gardening.

Slugs and snails are one of the garden's great destroyers. It seems that no sooner have you dug in your dahlias or planted your petunias than the small slimy ones destroy the lot. But how can you reduce the damage they cause without upsetting the natural balance in your garden, or running the risk of poisoning other wild creatures or even your pets? There is another way . . .

Contents

1

The natural solution to slugs and snails

Why use **natural methods**?

Walk into any garden centre and you'll see an array of slug pellets on sale, many of them labelled as being safe for pets and children. So why not use them? Why bother with natural methods of slug control? And what are 'natural' methods anyway?

Slugs and snails aren't the only ones that suffer when you put out slug pellets in the garden. Frogs, toads, hedgehogs and many birds like nothing better than a nice juicy slug – and if it's half dead as a result of eating poison then it's easy prey. Slug pellets work by disrupting the slug's gastric system, and most do this with a toxic chemical called metaldehyde. When the bird or animal eats the poisoned slug, the chemical can cause severe stomach

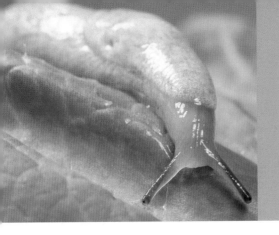

Working with nature, rather than against it, is the best long-term solution for controlling slugs and snails.

ache or worse – many people think it damages fertility in birds, for example. It can affect dogs and cats as well as wildlife.

What's more, some of the creatures that may be harmed by eating poisoned slugs are also slug predators – so you may not be doing your garden any favours in the long run.

There's another reason why pellets can be a somewhat short-sighted method of slug and snail control in the garden. Slugs and snails are attracted by the pellets, so you're inviting more of them – not to mention the whole extended family – into your garden. No one can guarantee that they won't come across one of your plants first, eat it, lay eggs and live happily ever after. It's just not that easy to outwit the slimy slug and snail.

Let **nature** do the work

No one can banish slugs and snails completely – if you want a living garden rather than a slab of concrete you'll always have to live with the enemy. But don't despair; allowing the forces of nature to work with you in the garden can pay dividends.

There are three main types of natural slug and snail prevention:

- Encouraging predators
- Looking after your garden
- Using traps and barriers

In this book we'll look at all three in turn. In reality you are likely to use a combination of these methods for effective control.

Encouraging predators such as toads (top), thrush (middle) and the hedgehog (bottom) is one approach to help you deal with slugs and snails.

Getting **started**

Banishing slugs and snails can seem a daunting task – but don't despair.
Remember:

- Standard (metaldehyde) pellets may seem like a 'quick fix', but they can cause long-term damage to your garden's natural balance.

- Non-chemical methods can be extremely effective, but don't expect instant results. Be patient!

- Get nature on your side; think before you plant. You may need to prepare the soil first, protect a plant until it becomes stronger, or grow something different.

How to use **this book**

- The book is divided into four main sections: an introduction to slugs and snails, followed by sections on the three main types of control.

- 'Action stations' at the end of each section provides a checklist of that section's main points.

- 'Resources' at the end of the book gives useful lists of suppliers and sources of additional information.

action stations

1 **Understand the enemy.** Get inside the head of the slug and snail! Knowing their likes and dislikes is the first step to slug and snail banishment. See pp. 14-20.

2 **Attract predators.** Slugs are a tasty meal for birds, insects and some animals – learn how to welcome these predators into your garden. See pp. 22-28.

3 **Care for your garden.** Stop your garden being a slug haven! Look after your soil, tidy up weeds and mess, and plant sensibly. See pp. 30-36.

4 **Set barriers and traps.** Stop snails and slugs in their tracks by using a barrier or trap, or by hand-picking (not for the squeamish!). There are many different types of traps and barriers available and you can make some yourself. See pp. 38-52.

2

Know your enemy

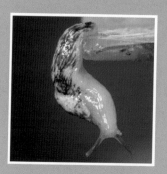

Understanding
slugs and snails

The first stage in any battle is to get to know your enemy. Understand the habits of slugs and snails, and which are the worst garden offenders, and you will be on the way to successful banishment.

As any gardener knows, slugs and snails adore dark, damp conditions. They're most active at night, and will be out in huge numbers just after rain. In fact, slugs and snails depend on moisture to survive and are at constant risk of dehydration. Their slime, which is mostly water, allows them to move along. So if they want to keep moving, they have to keep drinking – and eating your moist young plants.

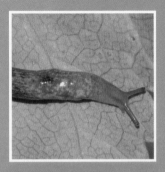

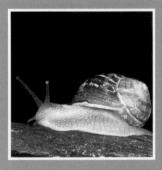

Getting to know your enemy, their habits and their habitats, is the first step to successful slug and snail banishment.

haven & hell

slugs & snails **love**	slugs & snails **hate**
Shade	Light
Darkness	Sunshine
Dampness	Drought
Cool (but not too cold)	Heat
Shade under stones	Acidity and alkaline
Overgrown hedges, ivy	Sharp stones and gravel
Plant pots	Copper
Damp wood and logs	Abrasive surfaces

Size isn't everything

Slugs attack plants both below and above the ground, while snails are surface scavengers. Slugs come in a range of colours and sizes, but it is the mini-slug varieties that you really want to look out for. Many of these live in the soil, come up at night, eat your plants and disappear. The largest garden slugs, by contrast, may only be chomping your compost or decaying vegetation.

Friend or **foe?**

Let's not forget that slugs and snails do an essential job: they eat dead plants, fungi and animal matter, and thus prevent the garden from becoming swamped in decaying matter. They also provide food for many wild birds and animals. We don't want to exterminate the whole population – just **keep them under control**, and encourage them to turn their attention elsewhere!

Not quite so guilty: The large and variable coloured black slug, does not cause as much damage as the smaller species. Likewise, the banded snail inflicts less damage to plants than the common garden snail.

fact file

- Slugs and snails are gastropods, which is Greek for 'stomach foot' – referring to the way they move along.

- There are 90,000 species of gastropod around the world – though many fortunately live in the sea.

- Mucous slime is a slug's or snail's 'weakest link' – they cannot move without it.

- A slug can eat up to forty times its own weight in a year.

- A slug or snail can have up to 27,000 tiny teeth-like protrusions.

- Slugs and snails travel at an average 0.007 miles per hour – but they're determined!

- Slug and snail trails contain the creatures' own scent, so they can find their way back home after dark.

- If you throw them over the fence they are likely to make their way back to your garden.

Rogues **gallery**

Slugs and snails are residents of every garden. Slime trails and damaged plants are signs that they are active. The most damaging culprits you are likely to encounter are the field slug, garden slug and the garden snail.

Garden snail. Shell marbled brown and black; approximately 40mm diameter. A familiar and widespread garden resident. Feeds on foliage.

Banded snails. Shells vary in colour combinations from white, yellow and grey to pinkish, usually with a darker band; approximately 21mm diameter. Banded snails are numerous in some areas but are *less damaging* to plants.

Also keep a lookout for smaller garden snails including the **strawberry snail** which is approximately 13mm diameter, has a flattened grey to brown shell, and is fairly widespread and destructive.

The garden snail (top) and the strawberry snail (above) are more destructive than the banded snail (p.16). Slugs (opposite) operate above and below ground and like snails are mainly active at night and in the wet.

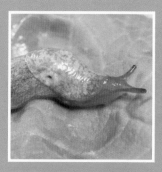

Grey field slug. Approximately 3–4cm long; *variable in colour*, (fawn/grey) with darker flecks or speckles. Attacks foliage, vegetables and some fruit. It will even feed in cool weather, above ground.

Large black slug. Up to 20cm long; black or orangey-brown. Often seen during the summer after rain. Although large, it is not as destructive as the field and garden slug, but is by no means innocent.

Keeled slugs. Up to to 6cm long; brown, grey or black. They are primarily soil-dwelling, attacking below and above the surface they will destroy foliage, roots, tubers and drilled seed.

Again, beware of smaller slug species including the roundback **garden slugs**. These slugs are no more than 3cm long; black with a pale stripe on the side and underside orange/yellow. They cause damage above and below ground and will attack stems at ground level and also burrow deep to attack roots. They love potatoes.

action stations

1 **Spot the difference.** Not all slugs and snails are as bad as each other; it pays to recognise the worst offenders.

2 **Know their habits.** Go out in the garden at dusk or after rain to find out where your slugs and snails are coming from.

3 **Recognise their habitats.** Get to know the conditions that slugs and snails love, and remove some of their favourite places (see page 15).

4 **Follow them home.** Gather them up and place them in the middle of your lawn. See where they go and how many birds spot them!

3

Encouraging allies

Attracting slug and snail **predators**

Slugs and snails play an important role in the garden's food chain, so why deny their natural enemies a tasty lunch? Make your garden a wildlife haven and nature will (almost) look after itself. From birds to beetles, you'll find there are lots of potential slug munchers out there.

Wild **birds**

If you have the space and facilities, chickens, bantams, ducks and geese will all feast readily on slugs. However, keeping chickens and other fowl is not a practical option for most of us – this is where wild birds can help. The long list of wild bird species known to eat slugs and snails includes blackbirds and thrushes, robins, starlings, rooks and crows, jays, seagulls and even owls. Song thrushes are particularly fond of snails. If you find a stone or paving slab with a scattering of broken snail shells around it, it's likely that a song thrush has been at work – the stone is used as an anvil to break the shell.

Attracting wild birds

Birds are always on the lookout for food, shelter and nesting sites – so a garden with a hedge or dense shrubbery is going to be a winner. Thrushes and starlings survive the winter by eating fruits and berries, so trees such as rowan, holly, berberis and crab apple are a good bet. Putting out nuts,

The thrush family are your allies when it comes to a war against slugs and snails. The song thrush and mistle thrush both thrive on snails and slugs, and the blackbird will eat a fair share of slugs too.

seeds and fat balls (you can buy these from most garden centres) is a good start, and nest boxes will entice birds to make a more permanent home in your garden. A bird bath encourages them to drop by for a dip and a drink – and if you have a pond even better. (For more on attracting wildlife see *Attract Wildlife* in the *Green Essentials* series.)

Hedgehogs and other mammals

Slugs provide a tasty meal for a range of mammals, but not all of them are creatures that you would want to encourage into your garden – the list includes badgers, foxes, moles, shrews and rats. Most people's favourite slug-eating animal is the hedgehog. This creature forages at night, hoovering up slugs with its versatile snout – it's particularly active on lawns and a hungry hedgehog can eat 120 slugs in an hour. Unfortunately, hedgehogs also eat useful garden creatures such as worms and beetles, but on balance they do more good than harm.

Attracting hedgehogs

You can encourage hedgehogs by leaving out at night a dish of protein-rich food – such as cat or dog food, or chicken leftovers – and a bowl of water (*not* bread or milk, which cause digestive problems). Once a hedgehog has taken up residency, reduce the amount of food you put out so they will be encouraged to eat a natural diet of slugs and snails. You can also provide them with sleeping areas in the form of leaves and suitable garden waste. (Always carefully move and inspect a pile of garden rubbish before burning it, in case there are hedgehogs sheltering inside.) You can even make a nest box for a hedgehog: build a wooden box (around 50 cm square), add an access hole and plastic ventilation pipe and part-fill the box with dried leaves. Tuck it away in a quiet corner, cover with soil, and leave undisturbed. If you're lucky a hedgehog will find it and breed more slug hunters in the spring.

The heavy artillery: hedgehogs, frogs and toads will seek and destroy slugs and snails in places you cannot reach.

Frogs and toads

Reptiles and amphibians may not be cute or cuddly, but they won't harm you or your plants – and they include several slug eaters. Frogs are the easiest to encourage; they hunt by ambushing anything edible that passes, and slugs frequently make up a quarter of their diet. Toads and slow-worms (a form of burrowing limbless lizard) eat slugs too, and can be helpful in drier parts of the garden.

Attracting frogs and toads

The best way to encourage frogs and toads is to create a garden pond, and this also has the advantage of attracting other slug eating creatures such as birds. Don't worry if your garden is fairly small – ponds do not have to be very big, and they get colonised by wildlife very quickly. The pond should be deep enough in the centre not to freeze solid in winter (around 60cm), and at least one third of the edge should be shallow and gently sloping to help wildlife move easily and provide an escape route for any creatures that fall in. A marshy area is also a good idea in order to provide shelter for wildlife, especially young amphibians. Before you start, though, it's best to consult a specialist guide, such as *Create Ponds* in the *Green Essentials* series.

Beetles and other insects

The garden teems with creepy-crawlies, many of which like nothing better than a slug and snail feast – and the same goes for their larvae. Slug-eating insects range from rare beauties like the glow-worm to common ground beetles and centipedes. Insects may not be as lovable as birds or hedgehogs, but they're worth their weight in gold when it comes to slug busting.

Attracting beetles and other insect predators

Planting a hedge or herbaceous border is a good way to attract insect predators. For further encouragement you can provide habitats and 'refuges' such as log piles, old wooden boards, slates and ponds. (These refuges also provide slug hiding-places; turn them over every few days and pick off the slugs and their eggs.)

Help from the light infantry: Ground beetles predate slugs and snails, while the lava of the glow worm feeds on smaller snails.

Biological control for slugs

As well as encouraging natural enemies into your garden, you can now buy in natural slug killers. These are microscopic nematode worms, that target slugs in the soil. They burrow into the slugs, releasing deadly bacteria. The slugs stop feeding, then die, releasing lots more nematodes to carry on the work. One application can protect plants for 6 weeks or so.

The nematodes, sold by mail order and in some garden centres as brands such as Nemaslug™, come mixed with a clay powder. This is mixed with water and watered on to the soil. To be effective the soil must be moist and warmer than 5C (10-25C is ideal). Late spring and early autumn are good application times. To protect potato tubers, apply the nematodes when the tiny tubers are just beginning to form (this usually coincides with the plants flowering). The nematodes are not suitable for use on heavy soils and are not effective against snails.

The microscopic nematode worm is a natural slug killer available by mail order and from some garden centres.

action stations

1 **Encourage birds.** Slugs provide a tasty meal for birds. Attract them with food and fresh water all year round.

2 **Add a pond.** Frogs and toads simply love slugs. Build a pond to entice amphibians into your garden.

3 **Call in the hedgehogs.** Put down some food and make a cosy nesting box – remember that they can eat 120 slugs an hour!

4 **Create habitats for slug busting insects.** Plant a herbaceous border or provide special hiding-places such as log piles, old wooden boards, slates and ponds.

5 **Find a nematode supplier.** These natural parasites can protect your garden from slugs for up to six weeks.

4

Gardens that slugs hate

Maintaining and planting your garden

Wildlife alone won't do all the work – you do have to put in a bit of effort yourself. The key is to make slugs' and snails' lives as difficult as possible by reducing their habitat. So look after the soil, tidy up, plant the right varieties and you'll be well on your way to natural slug and snail banishment.

Ground work

Slugs like heavy, wet soils, particularly for laying their eggs, to ensure that they do not dry out. They also like a surface with plenty of soil spaces to hide in. If the ground is covered with weeds as well, they think they are in paradise! So obviously it pays to eliminate these kinds of conditions.

Looking after **your soil**

Digging, raking and hoeing

A good digging session in early spring and between crops helps to destroy slugs and their eggs by exposing them to predators and the weather. Digging has the added advantage of destroying the cracks and clods that slugs love to shelter in. Raking and hoeing helps to disturb slugs' all-important slime trail as well as their hiding places, so try to rake and hoe throughout the season. Moving the soil surface with a rake in winter also helps by exposing slugs and their eggs to frost.

Drainage and soil conditioning

Wet ground should be drained and improved where possible by digging drainage trenches and adding gravel. Coarse grit dug into the soil can also help drainage. If you simply can't avoid wet ground try to avoid slug-sensitive plants in that area.

Raised beds

Raised beds can be a useful anti-slug tactic; the soil dries out more quickly, and you can create a mini-fortress by adding barriers such as copper tape. They also allow you to avoid heavy ground and to tailor the soil to meet your planting needs. Make one using wooden planks, or buy a kit from a mail-order supplier.

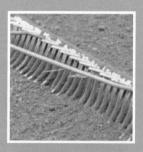

Improve wet soil with drainage trenches, gravel and coarse grit. Dig, rake and hoe to expose slugs and their eggs to predators and the weather. Use raised beds with a copper tape barrier in problem areas.

Weeding and tidying

Messy, overgrown areas are nothing short of a slug and snail hotel, with weeds providing food and shelter. If you can't keep up with the weeding try spreading a thick layer of mulch, preferably containing slug-repellent rough and jagged materials (see p. 43). And it goes without saying that dark, damp corners of the garden full of old flower pots and rotting logs are prime slug haunts, so have a clear-up and evict unwanted residents. But strike a balance, don't tidy the whole garden. Predators need houses too - leave some dark damp corners away from your vulnerable plants.

Plants and **planting**

One way to win in the battle against slugs and snails is simply to plant species that they don't like. Well-established, mature plants are usually more resistant, as are those with a strong aroma. Plants with rough, spiny, tough or hairy foliage and stems tend to be off the slug and snail menu too. But if you can't resist planting a slug meal put it in a sunny spot rather than a damp, shady corner.

Slug hotels: Dark, damp corners of the garden such as old flower pots, rotting logs and garden waste are prime slug haunts, seek them out and clear out unwanted residents.

not so tempting

Slugs and snails tend to avoid the following, though some may still be preyed upon when young and soft. When choosing salad plants go for those with red leaves, or peppery varieties such as rocket.

- **Vegetables and salads:** Onions, garlic, endives, lamb's lettuce, red lettuce, rocket, red cabbage, tomatoes, beetroot.

- **Fruit:** Alpine strawberries, tree fruit (apples, pears, etc.), bush fruit (berries, currants, etc.).

- **Flowers:** Busy-lizzies, begonias, perennial geraniums, foxgloves, fuschias, alyssum, daisies, evening primrose, snowdrops, fritillaries, daffodils, montbretia, hollyhocks, nasturtiums, heathers.

- **Herbs:** Marjoram, chives, mint, lemon balm, rosemary, lavender, lovage. Sometimes eaten: sage, thyme, fennel.

Only the **strong survive**

Slugs and snails, it seems, can spot a weak plant a mile off. Give your plants a sturdy start in life and don't let slugs and snails get the better of them.

- Sow or plant into warm soil – fast-growing plants are more likely to survive.

- If you grow seedlings to transplant, it's best to raise them off the ground in a protected area – ideally a greenhouse – before planting out. Use an appropriate multipurpose or potting compost to ensure sturdy plants. Harden off well before transplanting.

- When you plant out, put some distance between seedlings and slugs. Use barriers and hunt and remove tactics to try to create an 'exclusion zone'.

- Use traps to reduce the local slug population for a week or two *before* planting out.

- Plant decoys (such as some old lettuce leaves under a tile or brick) to divert slugs and snails from young and vulnerable plants. Young shoots are rich in water and nutrients, and so are especially attractive.

- Make a mini 'cloche' from a plastic bottle (see p.43) to protect newly transplanted plants.

favourite meals

Do you need reminding? Snails and slugs love:

- Beans, lettuce, spinach, cabbage, potatoes, coriander, strawberries, petunias, dahlias, marigolds, lupins, hostas, delphiniums, lilies, primroses … And more.

Variety and timing

Some varieties of the same plant are less appealing to slugs than others – so why make life difficult? It's also worth paying attention to timing: try to outwit slugs at their own game.

- Blue ribbed hostas seem to be less appealing to slugs than variegated ones.

- Potato varieties that show some resistance include Kestrel, Estima, Charlotte and Wilja.

- Lift potatoes earlier than usual to avoid the period of greatest slug damage in the autumn (this is when slugs like to burrow).

action stations

1 **Condition your soil and improve drainage.** Dig in the spring and rake regularly. Improve heavy, wet soil by adding gravel, sand, grit and well-rotted compost.

2 **Build a raised bed.** With the addition of a copper tape barrier these provide a good solution if you have heavy soil.

3 **Weed and tidy.** Deny slugs and snails their comfort zone – clear away mess and get rid of those weeds!

4 **Give your plants a sure start.** Grow seedlings indoors or in a greenhouse, use good compost, and protect young plants.

5 **Avoid slug favourites.** Hostas, marigolds and delphiniums provide an open invitation to slugs.

6 **Choose resistant varieties and plant varieties that slugs and snails hate.** Choose tough, rough, spiny or hairy plants, those with a strong aroma or plants that slugs hate (ask at your garden centre for further advice on slug-resisting plants).

Anti-slug tools
and tactics

Practical **action**

You've prepared the soil, planted sensibly, enticed the entire neighbourhood's wildlife to your garden – and still you find our slimy friends lurking about the cabbage patch. That's just the name of the slug and snail game, unfortunately. But don't despair: all wars use terrorist tactics and this one is no exception.

Pick and **despatch**

Hand-picking may not be for the squeamish, but it can be highly effective – after dark you can easily collect a couple of hundred snails and slugs in an hour. And if you place a decoy or slug trap on the ground (see pp. 40 and 46), the slugs will simply come to you!

Hand-picking slugs and snails can be highly effective and most productive by torch-light after sunset.

Hand to hand combat

The best time for hand picking is a couple of hours after sunset. Go out with a torch, and use gloves or tongs if the idea of picking up slugs and snails makes you shudder. You can usually find the greatest numbers on the lawn and pathways rather than on the soil itself. For maximum destruction you could even mow the lawn at night – but beware wandering amphibians and irate neighbours! Alternatively you can concentrate hand picking on vulnerable areas, such as a patch of newly planted lettuces. Clear a bed *before* planting by 'baiting' and hand picking.

Real slug haters will happily go on a search-and-destroy mission: tightly bind a hat pin or thick needle to the end of a stick, and spear them as you go just like a litter-picker in the park. Some people prefer to use a hand fork or sharp spade. **Alternatively, take them to a public field or common where birds feed; or to a piece of waste ground, cutting or embankment and set them free – keep well away from gardens!**

Disposal

The most 'humane' way of destroying your collection of slugs and snails is to crush them, but you can also drop them into a bucket of salty or hot water, or suffocate them by putting them in a plastic bag. Whatever you do, there's no point in just throwing them over the fence: slug trails contain their own scent so they can find their way home after dark.

Driven to **distraction**

Decoy plants and vegetation distract the slugs and snails from vulnerable plants such as new seedlings.

The decoy attracts the slugs, which you can then pick off.

Some decoy ideas:

- A week or two before sowing or planting out seedlings, put out old lettuce leaves covered with a roof tile to stop them drying out. Remove all the slugs you find lurking there before planting out your precious young plants.

- Also a week before sowing pile a heap of wilting comfrey leaves in the centre of the bed. Collect the slugs each day, and after a week remove the pile and dispose of it. After planting, ring the bed with more decoy comfrey leaves and replace them regularly. (Note: Comfrey leaves are also an excellent natural fertiliser. Chop them up and dig it in. It grows vigorously, so you'll never be short of it.)

- Sow or plant lettuces or french marigolds – a slug and snail favourite – around a bed a week or two before putting in your 'main' crop. Yellow mustard, cress or spinach also act as great decoys.

- The plastic bag method. This is half way between a decoy and a trap. Place a plastic bin liner or bag between your vulnerable plants, and put a couple of lettuces that have passed their sell-by date inside. Add a few handfuls of bran and a cup of beer. Leave overnight with the top open, to lure in the slugs. Gather up in the morning as they start to hide at the bottom of the bag.

Ready-made traps provide you with efficient hardware to help with the war against slugs and snails. For more on traps see page 46.

Slug and snail **barriers**

The idea of a barrier is to surround vulnerable plants with something that slugs and snails can't cross. This might be because the material is dry and scratchy, or because it contains a chemical they dislike. Barriers can never be foolproof and are vulnerable to wet weather, burrowing and slugs that hatch from eggs that are already in the soil.

All gardens are different, so it's really a matter of experimenting to see what works for you. Here's a run-down of tried and tested methods.

Bran is dry and difficult to cross, and as a bonus it makes slugs swell up when they eat it (making them great food for birds!). Also from the kitchen cupboard, try **coffee grounds and cinnamon** which are also disliked.

Copper reacts with the slug's slime, giving the animal a slight electric shock (it repels but does not kill them). **Self-adhesive copper tape,** available from garden centres and mail-order outlets, can be placed around the rim of pots; you can also buy copper mats to put under the pots. Alternatively, copper wire will do the job where tape or mats are not a practical option.

Petroleum jelly (Vaseline®), grease and lubricants such as WD-40® can be smeared around the rims of pots and other containers to stop slugs and snails climbing in; non-setting glue has the same effect.

Plastic is not the most attractive of materials, but feasible as a barrier in the vegetable patch. To protect small plants, make a mini cloche by cutting the bottom off a clear plastic drinks bottle, unscrewing the top, and popping it over; or make a 'collar' in the same way, but cutting just a 10cm section. For a whole bed, make a 'wall' of semi-rigid clear plastic; the sheet should sit at least 10cm deep and the wall should be at least 20cm high.

Sharp and spiky materials are difficult for slugs to cross; these include sand, sharp gravel or grit, crushed eggshells, coffee grounds, fir cones, mussel shells, pine needles, sweet chestnuts and shells of various nuts.

Wood ash and coal soot dry up the mucous glands that are necessary for a slug's movement; ash is also thought to be chemically repellent.

Raised beds can help protect against attack. Fortify with copper tape or wire affixed to the sides and place on a patio or over gravel.

Woodchips, wood shavings and sawdust may be too dry and fibrous for slugs to cross. Pile up about 2cm thick, away from the plant stem. Remember to replace if it rains, and check that the plants have enough water as these materials are very absorbent.

Commercial products. There are a number of natural products that are worth a try. These include **slug pellets based on ferric phosphate** (harmless to people, pets and wildlife - except slugs and snails); **copper coated fabric** which can be cut to size to fit around vulnerable plants; **barrier granules** that suck the slime from slugs and snails; **copper barriers;** and of course the slug killing **nematodes**. Try your local garden centre or contact the suppliers listed on page 54.

Garlic is the kiss of death for slugs. What gardeners have believed for centuries – that garlic deters slugs and snails - has now been proven by scientists. Researchers found that slugs were repelled by garlic and died within a few hours of being exposed to garlic oil. For now you could either plant garlic around crops or mix crushed cloves and water and spread near plants. In future expect commercial garlic oil sprays to be available.

barrier tips

Barriers can be an effective method of slug and snail control, but they do need careful monitoring.

- Take care to prevent earth and sawdust from becoming too wet.

- Make sure weeds don't grow over the barrier, giving slugs and snails a way in.

- Some substances are liable to blow away – check and keep them topped up.

- Don't mulch around small plants, or on a cold wet soil. Wait until the soil has warmed up and the plants are growing strongly.

- Remember barriers can trap slugs in as well as keep them out, so make sure the bed is as slug-free as possible before you begin.

Slug and snail **traps**

For anyone who's lost a lettuce too many there's something rather appealing about luring slugs and snails to an untimely end. You need many traps to cover a whole garden, though, so they're most effective when used to protect a specific area. The best solution is to combine traps with some of the other methods outlined in this book.

Which bait and when?

Beer is one of the best baits but you can also use milk or grape juice: the slug or snail is attracted by the smell of the liquid, falls in and drowns. It's a good idea to place traps early in the year (February or March) when there are few other food sources for the slug to investigate; thus you can catch the slugs that would otherwise have reproduced. The downside of traps is that you do have to empty them regularly – preferably every morning. Remember, the longer you leave it the more unpleasant it will be!

Beer traps

A beer trap is basically a plastic pot with a lid and some slots cut in the side. The lid is to prevent evaporation, to stop beneficial creatures such as frogs or beetles from falling in, and to prevent larger animals from drinking the beer. Sink the pot into the ground, but leave the rim about 2-3cm above the soil's surface – otherwise you'll end up drowning our valuable friend the ground beetle. You can buy ready-make plastic traps or make your own using a jar or yogurt pot – but don't forget the lid.

Other types of trap

Melon skins are an effective way to entice slugs, as are upturned grapefruit or orange halves that have been scooped out. Once the slugs or snails have collected you can dispose of them as outlined on p. 39. Or simply place wooden planks, tiles or upturned flower pots near vulnerable plants to provide hiding places (remember to leave a gap under pots for snails and slugs to crawl through), then pick the slugs and snails in the morning before the sun is too hot and the shelters are too dry.

A combination of good husbandry, barriers, traps and natural predators will help to ensure that damage suffered at the jaws of slugs and snails is minimised.

Summary

Keeping your garden entirely clear of slugs and snails is unrealistic; instead your aim should be to control the numbers by limiting the attractiveness of your garden for them and through physical controls. Once the numbers are reduced, your plants should be able to cope.

Slugs and snails
will steer clear of your garden if:

- You turn over all empty pots and containers in late winter and **dispose** of all the hibernating snails.

- **Rake out leaves and debris** from under hedges and the back of borders.

Removal and protection. Turn over empty pots and containers in late winter and dispose of all the hibernating snails. In spring protect new plants with plastic bottle cloches.

An empty yoghurt carton can be used to help protect against ground attack from slugs, snails and other pests.

- **Clear** any areas of abandoned wood, planks or bags – this is an ideal home.

- If you have a serious problem with slug or snail numbers it pays to go out at dusk every evening for a few weeks and **bag them** up. They can either be taken to some waste ground far enough away or placed in heavily salted or boiling water.

- Set regular **beer traps** sunk in the soil with about 1.5 – 2 cm still above ground (to prevent useful beetles falling in!). Scooped out halves of grapefruit or orange laid skin up near plants will also distract them. There are also many proprietary traps and barriers available from your garden centre.

- Use **copper tape** around the lip of pots – copper contains a small natural electric charge which deters slugs and snails – again available from your garden centre.

- **Baked eggshells**, sea shell mixes, coffee grounds, ash from wood fires and garlic juice are all unpleasant for snails or slugs to cross and can play a part in limiting their activities.

- For slugs – especially those who live in the soil – **nematodes** are a very useful biological way of controlling slugs especially when applied in early spring. Nematodes or eelworms can be purchased as a paste that you mix with water and water into the soil. The microscopic nematode (which has no ill effect on any other living things) enter the slugs bodies and fatally infect them with bacteria.

- Attract more wildlife and they will eat the slugs and snails for you – birds (especially thrushes), **frogs** and **toads**, **hedgehogs**, slow-worms and ground beetles all like nothing more than these juicy molluscs.

Beer traps, copper tape, nematodes and netting can all help in your battle to eradicate or prevent attack from slugs and snails.

action stations

1 **Search and destroy.** Go on a night-time raid to find as many slugs and snails as you can, and dispose of them effectively (ideally in a public field or on common ground where birds feed).

2 **Use decoys.** Protect your vulnerable plants by putting in 'suicide' decoys or rotting vegetation.

3 **Create barriers.** Surround plants with slug repellent materials such as sharp grit, ash or sawdust. Use copper wire or tape to protect individual plants, pots and raised beds.

4 **Set traps.** Buy a commercial slug trap or make your own simple beer trap – you may need several traps for effective control.

Want more organic gardening help?

Then join HDRA, the national charity for organic gardening, farming and food.

the organic organisation

As a member of HDRA you'll gain-
- free access to our Gardening Advisory Service
- access to our three gardens in Warwickshire, Kent and Essex and to 10 more gardens around the UK
- opportunities to attend courses and talks or visit other gardens on Organic Gardens Open Weekends
- discounts when ordering from the Organic Gardening Catalogue
- discounted membership of the Heritage Seed Library
- quarterly magazines full of useful information

You'll also be supporting-
- the conservation of heritage seeds
- an overseas organic advisory service to help small-scale farmers in the tropics
- Duchy Originals HDRA Organic Gardens for Schools
- HDRA Organic Food For All campaign
- research into organic agriculture

To join HDRA ring: 024 7630 3517
email: enquiries@hdra.org.uk
or visit our website: www.hdra.org.uk

Charity No. 298104

Resources

HDRA the organic organisation promoting organic gardening farming and food
www.hdra.org.uk
024 7630 3517

Soil Association the heart of organic food and farming
www.soilassociation.org
0117 929 0661

MAIL ORDER:

The Organic Gardening Catalogue
Nemaslug™, organic seeds, composts, raised beds, barriers, traps and other organic gardening sundries. All purchases help to fund the HDRA's charity work.
www.organiccatalogue.com
0845 1301304

Agralan Ltd Traps, copper tape
www.agralan.co.uk 01285 860015

Centre for Alternative Technology
Slug traps
www.cat.org.uk 01654 705950

Garland Products Ltd Traps
www.garlandproducts.com
01384 278256

Green Gardener Various slug repellent devices, **Nemaslug™**
www.greengardener.co.uk
01394 420087

Tamar Organics Various slug repellent devices
www.tamarorganics.co.uk
01822 834887

who, what, where, when and why organic?

for all the answers and tempting offers go to www.whyorganic.org

- Mouthwatering offers on organic produce
- Organic places to shop and stay across the UK
- Seasonal recipes from celebrity chefs
- Expert advice on your food and health
- Soil Association food club – join for just £1 a month

Soil Association
the heart of organic food & farming

Acknowledgements

Images

Becker Underwood, leading producer of nematodes (pp. 27, 51 and above).
Nemaslug™ is safe to children pets and wildlife and can be used throughout the garden, including on food crops.

Field slug *Deroceras reticulatum* – © **VCU Rothampsted Research** (pp. 6, 9, 19)

Black slug *Arion ater* – **A. J. Silverdale** University of Paisley (pp. 8, 19, 44)

Garland Products Ltd – Slug X traps (pp. 41, 47); raised beds (pp. 31, 43)

Various slug images (pp. 7, 10, 21, 24) – **John Robinson**

Images from **English Nature**: Toad (Paul Glendell), hedgehog (Charron Pugsley-Hill), thrush (Mike Hammell) (pp. 7, 10, 21, 24)

And thanks to Richard Jones and Laurie Brown for supplying various other images.